Agents Handbook
of
Black Bag Operations

Agents Handbook
of
Black Bag Operations

Desert Publications
P.O. Box 1751
El Dorado, AR 71731
501-862-2077

ISBN #:0-87947-212-X

First Printing 1978
Second Printing 1991

Desert Publication is a division of
The DELTA GROUP, Ltd.
Direct all inquires & orders to the above address.

Table of Contents

This Page Is
Intentionally Blank

Forward

Intelligence-gathering operations — more commonly known in the trade as "Black Bag" operations — have long been an area of great importance in developing or obtaining up-to-date information of relevent value on the opposing side, be it a military, political, or business enemy; in some cases, areas including matrimonial or a suspected business partner working a "deal" of his own with another rival business or for his own personal gain. In any of these, or other cases, the basic information — intelligence — is at the heart of the black bag operation.

The winner of any such match is, naturally, the one who has obtained or kept secret from the other side, the most accurate, up-to-date, and critical information about the opposite side.

Watergate is a prime example of this; it is also a prime example of what can go wrong when all your bases aren't covered from the beginning. It means that planning, research, some legwork, studying photographs and building plans, visiting the offices under a pretext to get more relevent information, or even pumping an insider for information (bribery is such a poor word, and one the law frowns upon) are necessary considerations.

Like Watergate, the worst thing that can happen is to get caught "in the act." The outcome of such an action is a foregone conclusion. Thus, it must and can be avoided; and this is done by using proper planning considerations and techniques. Even here an amateur can fall down on the job and leave some sort of evidence that may cause suspicion to fall directly or indirectly upon himself, or make the target wary enough to change designs, rewrite critical letters, or destroy future evidence you are looking for. Even this second method of partial compromise must be considered in developing the proper techniques for the successful Black Bag Operation. Thus, this book has been developed to detail the specific areas that are so necessary and vital to the success of the Black Bag Operator.

Throughout this book you will find, interspersed with the various criteria and techniques, ideas and techniques that can apply to various situations. They are included at these points since they refer to specific applications that can be used; to include them in a separate chapter would leave out the overall viewpoint that is necessary to understand them and their relation to use by the Black Bag Operative.

Introduction

Various definitions of "Black Bag" exist. Within the Intelligence and Security Community, they can include such diverse areas as:

— Courier;

— The depositing or extraction of people by covert or deceptive means;

— The insertion of special electronic and/or visual equipment for some form of monitoring into any controlled or limited access area in which the special agent's would normally be denied access;

— The entering and subsequent exiting of a person(s) into a denied area for the purpose of examining a portion or all of the area; the planting of hidden electronic (usually microphones), reording or relay equipment; document examination, or other equipment for use at a later time; or for the specific purpose of removal of certain documents or pieces of hardware.

Essentially, Black Bag is the term applied to any form of covert or surreptitious entry made into any area (building, apartment, house, business concern, foreigh entry) for the purpose of planting, inspecting, or removing something of value to the person (or his employer) performing the black bag job. In one form or another, almost all of the above definition areas are included in this generalized definition which is all-encompassing and intended to set out the parameters of this book.

The processes involved in Black Bag Operations can

include many specialized techniques and materials. This book sets forth the more acceptable types that can be necessary for the novice and intermediate individual performing such a mission. It also delineates certain features, technical materials, and certain exploratory techniques necessary for the successful accomplishment of Black Bag Operations.

* * * * * * * * *

The legality of surreptitious entries is highly questionable. In a FBI Memo, dated 7/19/66, an FBI official who administered Black Bag operations, expressed the opinion that they were "clearly illegal."

A FBI Memorandum made known to the Senate Select Committee to Study Government Operations with respect to Intelligence Activities, further discussed the policy for these type entries. The Memo, from W. C. Sullivan to C. D. DeLoach, subject; Black Bag Jobs, stated:

> *"We do not obtain authorization for "black bag" jobs . . . such a technique involves trespassing and is clearly illegal; therefore, it would be impossible to obtain any legal sanction for it. Despite this, "black bag" jobs have been used because they represent an invaluable technique. . ."*

Another Memorandum from the FBI to the Senate Select Committee estimated that from the 1960's until 1975, the FBI committed [at least] 491 surreptitious entries. [AUTHOR'S NOTE: How many more were made that are unknown may never be known and, of course, the FBI isn't about to talk about them.]

An internal White House memorandum, dated 7/70, page 3, was of the opinion that "Black Bag" jobs were illegal, but should be utilized nonetheless:

> *"Use of this technique is clearly illegal: it amounts to burglary. It is also highly risky . . . However, it is also the most fruitful tool and can produce the type of [information] which cannot be obtained in any other fashion."*

. . . .

"Surreptitious entry of facilities . . . can turn up information about identities, methods of operation, and other invaluable . . . information which is not otherwise obtainable."

At the outset, then, you can clearly see that even the Nation's most prestigious law enforcement agency and even the White House talk about their illegality, but are willing to perform them also. Since they are illegal, BE FOREWARNED!

This Page Is
Intentionally Blank

Operational Planning

Operational planning is, basically, detailed research which covers everything you may ever want to know about your target, and also a few things you really don't need to know, but should be aware of if certain factors suddenly change during the course of your Black Bag Operation. Essentially, it covers all facets of the proposed Black Bag operation that you can possibly obtain by any means.

The degree of success of a given operation is determined, to a great extent, upon the amount of planning that goes into it. If you put in ten hours of planning for every planned hour of actual operation, you are definitely much better off than if you visually observe an area for several hours, take a number of notes, and then go ahead that evening, assuming that what you have learned earlier that day will not change for the next several hours.

No rule of thumb can be given to cover every contingency and possibility that can crop up during a Black Bag operation, because every operation, no matter how similar to the previous or next one, is always different in one major or a series of minor ways.

PROPER PLANNING PREVENTS POOR PERFORMANCE

It can be taken for granted that a Black Bag job well planned is a thousand times better than an impromptu one. It is the climax of, possibly, hundreds of hours of research, detail, reading, and preliminary investigation into the target of the Black Bag job.

Currently, the field of Industrial Espionage and the attempt to obtain rival business secrets is one of the better areas wherein the Black Bag job can be used to a great degree of success. Although political and economic intelligence on persons, businesses, and groups rank in the same area, the basic techniques of the Black Bag job work essentially the same in the planning and preparation stage.

PLANNING

First, you consider the specific target to be "hit" with the Black Bag job. It might be an obscure office off the main road that has few, if any, street lights, no alarm system, very few pedestrians, virtually no night checks by the local police; or it might be the office of the manager of a major company that is located on the 20th floor of a 35 story building that is located in the heart of the business district and has numerous persons in the area from very early in the morning until extremely late at night, and has a series of roving police patrols, an external building alarm system, and an internal roving security force.

In determining the target, you are basing this on the type (as specific as possible, of course), quality and amount of information you are interested in obtaining. Obtaining the information may be by photographing it, removing it from the premises, doctoring it to alter the out-come, or merely placing a microphone in a strategic location so that you can overhear critical comments and conversations that have a bearing on what you are interested in.

Depending upon who you are, what your qualifications are, the amount of access you have, or can obtain, near and into selected area, will determine the amount of basic and detailed research and advance preparation you will have to do in order to perform a successful Black Bag operation.

Since this book must cover everyone, we will go on the premise you know nobody in the immediate target area who can help you — initially.

Thus, you start with your local governmental offices; after all, they are paid with your tax dollars, so let them be of great assistance in determining how much information they will willingly supply you. Obtain accurate detailed city maps of the immediate and surrounding area, in as much detail as possible. Get several copies. They will be used for various purposes. The larger the scale the better.

Next, attempt to obtain building plans. When a building owner/construction company applies for a building permit or permit to renovate a portion of the building, it is normal procedure for them to furnish a copy of the building plans or renovation plans in detail to the office that will furnish them with the legal authority or permit to do what they have set out to do. These building plans can provide a wealth of information that you should know.

Now comes the first research effort which will involve a little legwork, a lot of time, and some generalized drawing. Using one of the city maps, draw out in a larger scale the various city streets and blocks surrounding the general area in which you are interested. Use a scale of about 1" equals 100 yards, or something approximately to this scale. Tentatively mark out the known buildings, approximating their area in the immediate and surrounding blocks. Don't forget to include alley ways, open lots, building sites, etc. Everything that is known about the area will soon be indicated on this or another drawing.

Moving to the actual site area, you should roughly pace off the distances of the various buildings to make you drawing more accurate. Also note down the number of floors of all buildings. In your "wanderings" take note of the type of stores, their hours of operations, which ones have roving security patrols at night, where police call boxes are located, where there are telephone booths, and U.S. mail boxes. Also to be noted are other buildings that are higher than the one you are interested in and if people work there during the hours you wish to "hit" the selected site. If so, can they see the exterior windows of your building; and can they see inside. Also, you should be going down alleyways looking for dead end alleys, distances across the alleyways, rear heights of various buildings (not always the same as the front or street side height), loading docks, and legality of parking in the alleyways. Along the immediate street, note the movement of traffic, various parking/no parking areas, meter times and types of coins used in them. All this information will be indicated on your drawing or referred to in attached, detailed notes that clearly identifies what you are talking about.

Moving to the specific building in question, note the number of stories, approximate heigth of each one (to determine overall heigth), number and types of windows and how they open, and protrusions including exterior hooks and ledges for window washer apparatus, etc. Consider the ground, second floor, basement, parking sub-basements, etc., if any, and the roof areas. You now concentrate on them. On the ground floor, your observations should be made at different times of the day and night to get a feel for the amount of pedestrian traffic in and out of the building, whether or not there is a security guard arrangement, police patrols, number and type of janitorial services; number, location, and amount of lights; which doors are always unlocked and others which are always lcoked. Check out the loading dock, if any. See if there are a single or variety of trucks there and at what hours. Are the trucks unique in their appearance, or are there all types there at different hours. Are there trucks parked there overnight. Are the loading dock doors unlocked for janitorial personnel to empty trash through at night. These and other types of details are essential to know.

On the second floor, look to see if there are particular windows that are usually unlocked and open at night, or are all windows closed and locked at quitting time. What is the approximate distance from ground level to windows. Can windows be reached easier from alleyway or loading dock areas.

Drive into the parking garage in the basement/sub-basement, if possible, as a visitor to the building. Note number of cars at different hours, where they park, types of parking stickers used, if any, and the location of doors and elevators. Are the doors locked at night. Look for signs to this effect, and also check to see if the same applies to the elevator. Also, does the elevator automatically stop at the first floor at night, open and allow the security watchman to see who is going and coming into the building. Finally, does the guard use a check-in/check-out roster at night, recording the time and individual, and checking for identification. This portion can usually be checked by visually observing the front door area from across the street.

So far, this sounds like a lot of work — it is! But this is part of the comprehensive planning that must be considered for a real in-depth detailed plan that assures the successful Black Bag operation. Not every operation requires such a great amount of detailed planning, but you should know the proper steps to take should the occassion arise.

Now we move into the next phase and a slightly more difficult one (in some instances). If possible, you (or an accomplice) should use some form of excuse — very legitimate, if possible — to get yourself into the building during normal working hours, prowl the halls somewhat, noting down everything possible, and maybe, even into the specific office you are interested in. What you are looking for if further confirmation of some of your notes that were taken earlier: guards, security alarms, windows, lights, elevators, doors, janitors, etc. Also, if you can get yourself into the specific offices, you get a look at furniture placement, location of specific and secondary equipment, areas to avoid, internal secruity alarms, door types and locks, specific sizes and types of windows, overall area, lighting, and a general view of what can be seen out the windows (which indicates areas that can see in the windows, too). By being in the building at or near the end of the business day, you can observe any special security checkpoints, alarm settings, doors being secured and locked/alarmed for the night that might be there but not really noticed during the daytime since such procedures are not done then.

Within the offices proper, you can note the areas of importance, conference rooms, times for mail deliveries, and areas that will not be necessary to worry about. Attempt to obtain any brochures or other information about the business. They can be another type of valuable aid in further developing the scope of the operations.

All of the above can also help you in determining whether or not the operation will be during the day or at night. (NOTE: While "Black Bag" gives the connotation of a night job — black equalling darkness, many instances and occasions arise when the actual job is performed during the daylight hours, but it also means that the building is normally empty of personnel, such as during a Sunday or holiday.)

The next phase — PHASE TWO — now moves into high gear. Using the city maps, and taking the information from your notes and large scale drawing, color-code one map with the security arrangements of the area, to include disadvantages such as a neighboring building where the work goes on until extremely late at night. Indicate the local police precints, call boxes, lighted alleyways, parking restrictions, short-order (fast food) shops and restaurants that are open at the hours you consider working the Black Bag job.

Using another color, indicate 'safe' obstacles that can also assist you, such as telephone boothes, empty buildings, poor lighted areas, an all-night restaurant that might serve as a point for a lookout, city park, loading docks, etc.

Next, you consider the immediate surrounding buildings next to the one you wish to enter. Indicate whether the possibility exists of gaining entrance from the roof of one to another. Note the distances and estimate heigth factors. Later you will check these out ot see how feasible it actually is, whether there are roof doors or skylights available, roof alarms, specific distances, etc.

Using the building plans from the city planners' office of copies obtained elsewhere, select the floor of the one you want, the roof, the first and second floors, and the basements. You are rechecking the various entrance/exits, types of door locks (usually indicated in the specifications), etc. You will again, later if not already done so, check out as much of this as possible (or else have a accomplice do it for you).

Depending on how "close" to the subject area you need to be in order to acquire the necessary material for the Black Bag operation, other effective preliminary steps include:

— temporarily renting, or making a "serious" inquiry about renting space in the building. In this method, you are fishing for a lot of detailed information about the building, its occupants, security features, amount of traffic (day and night), etc., without opening yourself up for close scrutiny. Essentially, you are requiring information that will be specific, but seem like a normal request to the owner/manager. Some considerations can include getting an up-to-date copy of the floor plans (especially if the building is somewhat old and there have been numerous renovations, or if the building has open 'bays' with semi -permanent walls and false ceilings throughout (this allows owners to reconfigure to suit new oppupants with a minimal cost outlay). The questions about night and day security, provisions for entering the building at night/weekends/holidays, janitorial crew and their hours (they won't disturb me if I need to work late at night will they?) Sub-basement parking arrangements; elevator night keys; frieght elevator size for moving in your office equipment, etc. The type of businesses within the building can also lead to other Black Bag jobs, so inquire as to the other businesses as well, how long they have been there, etc. There is never too much information that you can glean from such conversations with individuals if they think they will profit from questions; especially an owner anxious to rent!

Another possibility to seriously consider, depending upon the circumstances, is to rent space in an adjoining building. If this is done, it must be accomplished with several things in mind: (1) it should be on the same floor or at least one floor above that of your adversary; (2) it should be opposite the proposed office (as much as possible; not necessarily absolutely and exactly opposite since you can better view the interior of the office from a slight angle); and (3) your actions within the temporary office rental site should either be that of a normal small businessman setting up an office, or you should pull shades and possibly put up some miscellaneous curtains, etc., to make it seem like a business office that does not want people peering in. From these offices you can then observe the daily activity, individual persons, and possibly note certain movements, location of critical items, the possibility of hidden silent alarm switches, etc. One other point on this to consider is that if you believe the office has master safe, it is usally located in an 'inner' office, i.e., one away from the main entrance and usually located in another office, which could conceivably be one with a window facing your rented office windows. If so, you have the possibility to observe, through a set of binoculars or small telescope, the opening of the safe, thus ensuring that you have the proper safe conbination. Also, you can note if they make telephone calls just before opening the safe every day; this means there is asilent alarm system hooked up to the safe.

The above method is sort of a last resort, since it is somewhat expensive and very time-consuming. The first and the following methods are preferred and can provide much better results, but of course, your particular needs and requirements dictate what method(s) you will use.

The final method I wish to mention is that of a temporary hire employee with the building management. In such instances as this, you need a certain amount of background 'papering', that is, the creation of a previous job employment history, living locale, etc. I prefer to think of this situation as one wherein the individual, usually an accomplice, seeks employment as a new arrival in the city, literally willing to take on any type of semi- or permanent work, starting at the bottom of the ladder — as a janitor of part of the building work force that might be in a position to paint, renovate offices, etc. In such an instance, the person would use a false name, background, and give a fake Social Security Number. In considering his cover story for this purpose, he should seek temporary rooming space at the local "Y" or some hotel that does not ask too many questions or be specific about identification. As for the employer, you should count on not more than two weeks maximum working for him. By the end of two weeks he will know whether or not you really have any qualifications for the position, or become suspicious as to why you would want such a job when you seem to have much potential. Also, very much longer and the money he withdraws from your paycheck to pay into the employer's contribution portion of the Federal and State taxes will return stating that either the name, SSN, or account belongs to someone else; then you will have several questions to answer, and your employer probably won't believe any of your answers.

In your position as a handyman or janitor, you should have a certain amount of freedom to roam and acquaint yourself with the building layout. Thus, you can observe the building at all hours of the day and early evening, or on into the night if you are a night-time janitor. As a janitor, you can check out the trash from the office(s) and pick up general and miscellaneous literature, imcoming and outgoing letters and carbons, and get a better feel for the people in the offices concerned. You also have the opportunity to closely check out the area for alarm systems, open and silent types, door locks, key types and numbers, and if the locks are masterkeyed. As a janitor you see all offices without anyone really becoming overly suspicious of you.

Depending upon your ability you can get hold of several different keys and, using your key impression pad, make a good imprint of each, and from them cut your own key. Of course, if you have the opportunity to hang onto the keys overnight, you can have a duplicate made or make one yourself. If you can locate a broken door, say in the repair shop or building storage area, remove the lock (only temporarily — overnight), dismantle it at home and determine if there is a master key. If so, you should go ahead and cut a master key for it, and the next day replace the lock on the broken door. As the same time, you would have the opportunity to test the key out in several locks; this assures you that the master you made is still in service.

Not withstanding the above procedures, you should still check out the building watchman or security guard force, if any. Using various ruses, through an accomplice, you can get an 'inside' view of the security system in the building. I prefer to think of myself as a free-lance photographer/writer working up a story for the local newspaper, and use the guard/guard force personnel as the subject of the story, telling about their work, with appropriate pictures, etc. This would entail you taking the nighttime tour of the building with them as they make their rounds, a little background of the guards and a lot of ego-building of them on your (or your accomplice's part) in order to put them at ease and make it seem that they are really important.

Another angle to consider, although it is somewhat questionable, is that of a freelance writer/photographer who has been commissioned by X magazine (low-grade pulp magazine) to prepare a number of pictures for an upcoming story about an office girl attacked by the night watch-man/night security guard. Thus, you are in a position to have a second accomplice, a fairly good looking young girl (not overly beautiful — remember that these magazines don't use top-notch models in their pictures) who is willing to assume certain poses in conjunction with your work. Naturally, she is nearby. To make it seem more legitimate, produce a 'standard printed release form' for the guard to sign with his name, address and social security number. The form should indicate that for his being in the pictures, he would receive a sum of $50-$100 when the pictures are published. Note this phrase. . .of course, they will never be published, but you will have some interesting pictures of him that you can always hold over him at a later time, if need be. Essentially, you are conducting a business deal with him, without his employer's knowing it . . ."a little ready extra cash for the underpaid security guard."

Since you can possibly be seen form the street, and the model in the picture is supposed to be an office worker, suggest that you adjourn to an empty office where the pictures can be taken, and inquire (suggest strongly) that they be on another floor (naturally, the office you are interested in will do just fine!) thus moving you elsewhere — from public view. Encourage the guard slightly, point out that realism — to some extent — on his part is fine, but don't overdo it. . .it's just a posed picture. Let him know that the pictures will be taken with his back to the camera, so he won't have to worry about being recognized. While you are taking a number of test pictures, resetting lights, and also the camera angles, get several pictures of him and the girl together in possibly compromising poses, and ensure that the calendar date and possibly a clock are in the background, thus ensuring that you have positive proof that he was not doing his assigned guard job at a certain time on a certain date. Of course, you may want to use a miniature camera for this work, or just indicate to him that you are taking light meter readings with the flash unit on your camera being activated at the same time to ensure that the intensity in the room is not too high with the flash unit in operation.

While such a technique may seem 'absurd' or 'going overboard,' remember that this might only be one of the many gambits you are able to use in order to get the necessary information for the Black Bag job; at the same time, you are getting an unpaid and unwitting accomplice to help you learn about the office(s) in question, and more about the security system of the building itself.

Using the examples above, the information gleaned and the photos taken can assist you in filling in the gaps about various areas of the office proper, such as: exactly which and who's desks are where and most important; where certain files/file cabinets are located (especially if there are lot of them); distances; light switches; windows that will need blackout curtains and approximate sizes; identification of areas/rooms where you may want to insert miniature listening devices (the word 'bug' seems so unprofessional at such a time as this); and the types of locks on the doors, desks, safes, windows, etc.

There is never too much information gleaned from your research; you should know the area as well as someone who works there on a daily basis.

I mentioned earlier contacts with janitorial personnel or else inserting yourself or an accomplice as a temporary janitor. This opportunity presents and allows you to obtain other valuable information, such as the daily trash pickings. The initial thrust of the Black Bag operation for a particular business/office might be shunted aside, or provide certain information through the trash that can provide the answer(s) you are looking for, or provide other areas (within the same business or another business) for which you can obtain further information. Also, it can lead you to other business concerns that may provide related additional information, or even new areas to prove that may be more lucrative sources of income for you.

A long shot, and somewhat more dangerous in developing

initial information about the business is to take the building manager or night watchman into your 'confidence', using a set of phoney credentials. Since they probably will not know what the real ones look like (Have you ever seen a set of Secret Service or FBI or other government Special Agent credentials before? Probably not. Well, two to one neither has the building manager.) You will probably be accepted without a further check. But just in case, have a professional-looking identification calling card made up, with a temporary telephone number on it, such as a flop-house or corner telephone in a quiet section of town listed. Naturally, you will have another person (accomplice) answer the telephone, answer various questions that may arise with a professional tone, and thus 'verify' your legality with the building individual. At this point, the accomplice might ask to speak to you personally, and your side of the conversation would indicate other official type 'business' that might be ongoing in your office, referring to this and other operations in the works, mentioning the possibility of getting higher headquarters involved since it is a sticky situation, and you might even require assistance of the local police in an upcoming raid, etc. This technique -- the impressing of individuals — can be used safely only a couple of times before too many people become aware of it and move in on you, or start checking up on you without your knowing about it.

If you obtain access almost immediately, request the individual to leave, since you are on 'official' business' and they do not have 'a need to know' exactly what you are doing (but indicate that they should check the office immediately after you finish to make sure you haven't disturbed anything; this is also a double-check for your own self--security, also. But let them know they should stand by outside or closely nearby as a lookout to make sure you are not interrupted while conducting your 'official business.' Thus you have a ready and willing accomplice to keep the night janitorial personnel out of the office and also to let you know if someone else comes into the area. They become, unwittingly, a compromised accomplice to the Black Bag job and are your lookout (at no cost to you), believing they are working in the National interest (oh, how often we have heard that phrase . . . and it still works)!

At a time such as this, you may also wish to have a key made from the one used to open the various office doors. Use your impression pad and also note down the key number and type of keyway used. This can come in handy later on, especially if it is a master key, and you consider the possibility of doing a Black Bag job on other offices in the same building.

This Page Is
Intentionally Blank

Plan of Entry

Next, to the actual Black Bag operation itself, the most critical phase of the overall operation is the plan of entry. This is critical since it:

— Must develop all protective measures for:

(1) personnel taken into confidence

(2) personnel actually performing the Black Bag operation

(3) personnel being used as lookouts and for counter-surveillance

— Develops the areas for the actual entry and exiting from the operational site

Some of the information herein may seem somewhat repititious as it seems to be have been touched upon in the previous chapter, but in reality, it is for another purpose. I believe you will recognize the need for certain other measures that will involve (possibly) some of the same individuals and techniques that were originally used in the basic operational planning phase.

This section is broken down into two parts which are discussed below.

THE PRELIMINARY PLAN.

If the pre-planning stages have all gone according to plan, then much of your basic groundwork is in hand, but it needs to be finalized into a workable format. Essentially, this plan — when in its final form — is tested by an actual survey, done at approximately the same hour, and within the time constraints of the actual Black Bag operation. It does not attempt to actually enter the site, but more-or-less check out all of the acquired information, such as checking out street lights, general and specific security problems, etc. It confirms by actual visual observation these details (as much as possible), noting in detail all times, movements, positioning (to include prepositioning personnel and/or equipment), the various details and who handles each portion; ensuring that the escape route is clear and vehicles (if used) are in excellent working order.

Depending upon the amount of necessary detail required, it is at this point that you may move to develop a scale model of the actual site in order to better understand the various details, who does what, who is located where, line of sight, etc.

PRELIMINARY CONTACTS.

It is this phase that somewhat duplicates, in part, the obtaining of various contacts within the selected operational site. When the site has been designated, initial contacts are almost a certainty. This becomes necessary to:

— Avoid contact with the office and/or personnel working therein;

— To ascertain any other type of information which you may not be able to gather yourself because of varying circumstances.

In developing these initial contacts, whether or not they are the same ones you used in the first phase (operational planning), it is not enough to seek out the individual who may need extra cash, have a personal grudge, etc., but more than this, you must ascertain first of all that the person will not become a 'double-dealer' and work with you while, in reality, he is working for management — AGAINST YOU! His personal characteristics, amount of loyalty to his employer or the building manager/owner, etc., muchless his ability to keep a secret, his potential belief that you are someone else (and this belief cannot be shaken, for you must prove or have available appropriate credentials to 'prove' you are who you claim to be . . . to his complete satisfaction), he does not become overly intoxicated on the job, etc. Only after you, or an accomplice (better him than you) have ascertained these facts, and are completely satisfied, should any given individual be approached. It may be difficult to persuade the individual(s) to cooperate, especially when there may be nothing in it for them, or they question the need for such secrecy. You must 'sell' them on the idea that it is in their best interest to assist you in any manner possible, without desclosing to them the true purpose of what you are really intending to do. Reasonable and quite plausible excuses can be developed to fit a variety of situations; and you should always have a backup story, that seems more plausible AND OFFICIAL-SOUNDING, than the initial one just in case they are slightly suspicious.

The impression made upon the individual during the initial contact/recruitment meeting is of the utmost importance and its value cannot be underestimated, for it must create and instill a confidence and manner of officialdom. The individual must be convinced, beyond any doubt, that what you are doing is in the best interests [of the government, official duty, partiotic duty, etc.] and that the people/person who will be performing the actual 'entry' (Black Bag job, but don't tell him this), are highly competent and professional, and that there will be no risk of discovery or future risk to him.

Remember that Watergate was performed by professionals, but then they didn't take the building security guard — who made the roving patrols of various offices — into their confidence, even with a false story. This was their downfall . . . and that of an Admimistration!

By intimation, leave the impression that unless this 'entry' is performed and the individual's attitude might well be considered as being against the best interests of . . . (local police, federal government, or whomever you claim to represent). If the individual's cooperation is not wholehearted, he may be censured by his employer, etc. Other ploys can also be introduced to show that a 'black mark' will be put on his file "in Washington, or whereever," indicating that he is unwilling to do his duty to assist the proper authorities in carrying out their duties. This has a great affect on many people, and they ultimately believe that they must help.

In nearly every case, there is no 'reward' but only the secret satisfaction for the individual that they are doing something in the 'national' interest . . . . You can go so far as to indicate that depending upon the circumstances and what is obtained, your superiors will seriously consider an incentive gift be made to the individual for services rendered. Use this as a persuasion tactic, to be given after the 'entry' is performed. The trouble with cash persuasion is that a person's loyalty cannot be bought and thus, it can become a backfiring technique in which you can become trapped at the last moment. If used properly, though, you could offer a gift of nominal value to him several days after the entry was accomplished, assuming that the premises in question did not suspect something and contact the internal security force or local authorities. On the other hand, you could just forget the individual and the incentive gift, and hope that he would assume that either the information obtained was not worth enough to allow you to offer him something, or else that he had been shafted — in which case he probably will keep his mouth shut anyway. In essence, it is very dangerous to put cash in hand to the individual before the entry (since it may seem a bribe, and thus, illegal in more ways than one), and far more dangerous to give it to him after the entry since he may be watched or under general suspicion.

Right here brings up a point that shows it is sometimes good to have an accomplice doing your initial contacts, etc. First, you are never recognized and cannot be identified. Second, you do not need a disguise in order to casually frequent the area in question either before or after the entry has been made.

Once the individual is totally on your side, he can be of immense assistance in helping you to neutralize other building employees, employees within the office(s) you wish to enter, and in developing specific information (otherwise unobtainable) on internal security practices and procedures. He is more than likely to know the habits, workhours, specific areas in which certain individuals work, and the movements of various individuals who work evenings and weekends. This can be of help to you on the day/evening of the Black Bag operation, since you will know (roughly or specifically) who and what areas to avoid, and specific times that are available to you when performing the entry. Many of the larger buildings, nowadays, have a night/weekend security guard on the main doors, and a alarm control board which monitors all other lower exit doors (but not, necessarily, the roof door). But, since the night shift crew (janitors, etc.) need to deposit trash, etc., in the outside trash receptacles, there is usually one door that is not alarmed during all hours of the evening and can be opened from the outside with relative ease, if need be.

The log-in/log-out register for the building, kept at the security guard's desk, will reveal the amount of individual traffic in any/all portions of the building, which offices are occupied, and give general indications of the number of people working. A fine point here to keep in mind is that if there are no people working that night or just at certain hours, and the night shift janitorial personnel have completed their jobs and departed, the elevators will not be in use. Since the main elevators are near the front entrance/guard's desk, he can easily ascertain whether someone is using then when no one is supposed to be in the building . . . BE FOREWARNED. Remember this and coordinate accordingly. If the guard makes nightly rounds, get a contact to work out the times he is at specific points of the building, especially near or in the area in which you will be working. The watchman rounds are of paramount importance since over a period of time, he may start to vary the rounds to break up the monotony. Even this can become a ritual, and sometimes this variation takes on a pattern, i.e., once every fourth night he performs the nightly rounds in reverse; by reversing the time/distances between various points, you can thus determine where he will be at any given point in time, and thus ensure your safety and security one step further. This is of importance when planning your entry or exit from the site; you certainly don't want to have successful entry and then run right into the security guard when preparing to exit from the site; on the other hand, you don't want to run into him just as you enter the building.

BUILDING SECURITY – SPECIFICS

In preparation for the actual entry, everything possible should be covered in order to make the entire entry (and exit) a smooth procedure, thusly easy. When possible, and if obtainable, pass keys, either to individual doors, or for the entire building shoud be obtained, either by surreptitious means (swiping one, making copy, and replacing it), or by impressioning one on a key impression pad at an opportune moment. Sometimes, you may require several 'masters,' since larger building complexes, have masters that cover the outside doors, all inside doors, or just all doors to be a specific section or the building or individual floor or office suites. When the keys cannot be obtained, you will have to know about the types of various locks you will encounter, the types of keyways they use, and sometimes the key numbers (if available) that can assist you in cutting a key for the lock. Extra sets of keys should be made, one for the individual making the intial entry into the building, one for the last person in/out, and one or two others sets for outside people, should they need them. [Note: Sometimes various master keys come in handy later on, since in many cases, only the locks to specific areas are changed on a random or specific schedule, but the overall master key may well remain the same; this means that you can effect re-entry into the building site at a later occasion for another Black Bag job, if necessary.]

I mentioned in the previous chapter the fact that detailed (as much as possible) floor plans should be obtained. Now is the time to bring these up to date, to take into consideration any changes that have been made since your intial building site survey. Particular attention should be paid to:

- The Office and immediate surrounding offices (if any)
- Hallways and stairwells (for quick escape routes)
- Elevator(s) (number, where located, and relation to guardposts, and exit doors on ground floors)
- Second story of building (if it is deemed necessary to exit from second story and disturb the guard)
- Roof (for possible entry/exit)
- Types and sizes of windows and their latches (for entry/exit, and also for sizing of blackout curtains)

By this time, you and individuals accompanying you on the Black Bag job should be totally familiar with the specific area of offices for which the entry is being made. Only further study on your part will produce other specifics of which you should be aware.

Within the hallways, stairwells, and janitorial closets opening onto the hallways, look at obstacles and protrusions that can temporarily hide you and/or your partners.

Note the size of protusions, and what can be seen from either side of them, how deep are they, etc.

In the stairwells, can you see down more than one flight of stairs, and what is the amount of light in the stairwells. Are shadows cast. Can movement be heard; do the stairwells 'echo' movement.

In the small closets, how much space is available, and are there various janitorial carts, etc., that can provide line-of-sight protection from a quick view of the area by a passing security guard or other individual.

Also check out trash storage areas. In this respect, where is trash stored on separate floors before it is taken to the main trash area for removal from the building. Is there a possibility that you can effect an exit from the building by hiding in boxes/bins of trash. Also, the trash areas at the rear of the building, or side thereof. Recall the possibility that the doors here would not always be alarmed. This is an area to consider for entry/exit, assuming that a guard is not on duty at this point when trash is being moved outside.

A point to consider indirectly to the exiting operation is the location of mail boxes/mail chutes. In some cases, the information, be it actual documents or photographed material, can be slipped into a pre-addressed stamped envelope(s) and conveniently mailed to yourself at a P.O. Box or at another address of your own choosing. Thus, you carry out no incriminating material that would compromise what you were in the building for. Using the film mailers, you could also take the lenses off the camera and mail them also, carrying out only a stripped down camera, or if the camera is expendable, dropping it in with the general trash, but only if you can be assured that the trash will be hauled away within the next 12 — 18 hours.

At this point we will discuss further the Preliminary Space Survey in more detial. A preliminary survery was made when you first slelcted the site and got as many details of the specific offices.

This portion of the security inspection is vital, as it is one of the last steps you take in order to assure a smooth and successful Black Bag operation. Attention to detail is very relevent.

Even your inside 'contact,' if any, must not know the real reason for this survey should he accompany you to the proposed site for 'planning' purposes. This check should be made under certain conditions, which as closely as possible, duplicate the time you will be making the actual entry. Note that in many instances this survey will not be possible, thus to some extent, you will be working in the dark. But should it be possible, it is to be done in silence and with as much caution as possible. In this instance, the silence is maintained until you have a chance to check for hidden security alarm devices, microphones, and possible hidden cameras or other electronic light/sound security devices. Note the initial entry way with extreme caution. Look just inside the door, above and behind, and at the opposing walls for light-activated relay unit, and also for concealed alarm trip switches or 'safety' switches which would turn off the alarm. Some trip alarms have a 30 — 90 second hold device, allowing the person entering time to walk to an unobtrusive point and turn off the alarm before it sounds.

Fortunately, most major buildings that have internal security guard forces do not rely upon interior alarms. But in instances wherein the material or offices hold items such as industrial secrets, Research & Develpment plans and products, expensive cameras, electronic equipment, or large sums of money, the room or office may have a local alarm present.

Once in the room and having ascertained that there are no alarms or that the alarms have been deactivated, you should look at the windows; measure them for the size of the blackout curtians and the number required.

Next, move to the various desks and cabinets, noting the type of lock and keyways. Look in the manager's office or the supply room for a possible key storage container that might have extra keys or master keys to the offices. These containers have a small, not-too-secure lock on them, and are easy to pick open; in many instances, by rote, they are never even locked, but merely closed at the end of each workday.

If there are safes in the room with combination locks, you should carefully check out the manager's office plus the desks/offices where the safes are located. Many times the combinations are hidden 'in plain sight.' Look at the calendar pads, telephone lists, designs, doodles, etc. These oft--times conceal the combinations in various ways. Some safes have a key and a conbination lock on them; if so, again check the desks and the key control box for the key. Keys may be just dropped in the desk drawer, or taped under the drawers, out of view. Make duplicates on the key impressioning pad, noting down the keyway and number for future reference.

The final portion of the survey is that of the exit route from the building. From the office(s) proper, determine the most secure (not necessarily the quickest) route out of the building that would avoid security guards, alarms, other personnel working the building, and note time it would take to walk/run from the offices to the outside of the building. TIMING IS IMPORTANT. Observe how the windows open, and if it is possible that the windows lock, automatically or with a little help after closing. Check the possiblility of roof exit to another roof. Check out the side doors, delivery and storage door exits, and the lower garage exits, if there is one.

You should mentally notice if you will have problems with your equipment carrying cases (if any). There may be a problem with size or the number of them, or overall weight when it relates to the timing or distance that must be covered. Whenever possible, make sure the exit route is different from the entrance route. Also, keep one or more alternate routes in mind.

Outside the building, you should take note where your lookouts (if any) will be stationed, and if one or more of them will be assisting you once you physically exit from the building. How far away from the building are they, and can they move a vehicle very close to the exit point to immediately pick you up for immediate exiting from the operational site are two very important questions you must ask yourself.

Operational Considerations: Entrances & Exits

The actual entry and exiting from the building site should be performed in total silence, as much as possible, with the following firmly in mind, and practiced by all, whenever possible:

— Two-way radio communications should be the responsibility of only one person during these portions of the operation. The communications expert should be in constant contact, but maintain radio silence, except in an emergency, when only code words will be used. The individual carrying the communications equipment into/out of the building site should have and earpiece and a clip mike (if possible) with a cut-off switch on it.

— The individual leading, and the one bringing up the rear, should have a complete set of keys for opening and relocking all doors. The individual leading should have ready a set of appropriate lockpicks, if the need for them arises.

— Equipment should be packed in lightweight containers, and carried by the individuals who will be using them. It is possible, though, that all equipment for each person may be carried on conveniently-prepared pockets on their clothing. This makes movement much quicker.

— Each person should have a one-way radio link to the communications center in order to hear, immediately, any vital message that may effect the entire operation, and thus be able to alert the others in case one or more have removed their earpieces in order to better do their job (such as the safe expert who would probably remove his earpiece in order to use the electronic stethoscope in determining a safe combination).

— Either the lead or second person should carry the blackout curtains and fastening tape.

— Either the second or third person should carry the Emergency Interruption Kit for subduing and restraining anyone who might wander into the immdeiate area and must be temporarily put out of commission until after the operation.

— Each person should wear a close-fitting face mask when necessary. During the period getting to and after leaving from the site, the masks are removed.

— If security alarms are known, and they can be put out of commission, then the alarm expert should follow immediately the individual opening all doors, or in some cases, precede him to affect neutralization of all alarm systems.

* * * * * * * * *

After the exit, the getaway from the immediate and surrounding area is a vital must. It may be affected by auto, foot, or even by bicycle. Don't laugh just yet. If the operation is performed on a weedend, who would look at several bicyclists with packsacks who are evidently on their way out into the country for a long bike ride and camping out overnight? not many!

Upon exiting from the site, you should be prepared, at the earliest possible moment, to change clothes, or have other clothes underneath your operational clothing, to affect a complete change, which might also include a 'personality' change. In the instance of a bicyclist, wearing a pair of shorts, and having a small backpack available to put all your other clothes into would not be a bad idea. Also, you might want to have a set of goggles and a small bicyclists' helment available to wear. Be sure your shoes and all attire match and seem plausible to the casual onlooker.

At night time, depending upon the hour, you may want to change to a more formal evening dress, deposit the other clothes and equipment cases in the trunk of a car, and depart for a somewhat distant restaurant or nightclub to "wile" away the hours and also create a more formal alibi for everyone concerned. After all, who checks exact times in a nightclub, especially when you are spending enough, but not too much, money there.

This Page Is
Intentionally Blank

Post Operation Actions

Once the Black Bag operation you have worked so long and carefully on, has been sucessfully executed, you have a number of options to consider, and certain ones that must be executed almost immediately.

First, the imformation, be it film or actual documents, and any/all peripheral and associated information obtained must be disposed of. If you are under contract, or doing the work by yourself, then planning on hunting up a buyer, will determine what you will actually do. In any case, get rid of the information; stash it away not at home or at a friend's house. I always liked the woods, preferably a state park You can drive in any time, take a short walk, and having the material properly wrapped to prevent mildew from dampness and/or direct rain, bury it relatively near a campsite, or off a wilderness trail. You don't carry a shovel, but you do prepare in advance with a walking stick with a small metal cap covering a pointed tip, which you will use for digging the hole.

If you already have a buyer, arrange to meet him some distance away, such as another city, or else rent a bus station storage locker, meet the contact at an intermediate point, and swap the key for your part of the bargain — the cash! If you're really sneaky, and want to keep a step ahead of him, you will get photographs (back view) of you (unrecognizable) putting the info into the luggage locker (keep one identifiable item on top) [MAKE SURE NO FINGERPRINTS OF YOURS ARE ON THE LOCKER, OR THE CONTAINER HOLDING THE DOCUMENTS]. Also, you have someone standing by when it is picked up, and with a short-range lens camera, photograph the person, several times, walking up, opening, removing the package, and of course, seeing a document on top, putting it out of sight. [There is nothing like a little extra insurance against your 'sponsor' just in case you get into a jam and need a lawyer or bail money; he's always willing to be of quick assistance, muchless a good character reference, if need be.]

Now, if you mailed the material to yourself at a P.O. Box or other address, you can safely let it sit around for several days before picking it up; it may be several days, in fact, before it arrives, so you can probably relax for about a week until the items are to be picked up either by you or your 'sponsor.'

Radios, police/private security frequency scanner units, one-way radio units, and other electronic devices developed specifically for a given Black Bag job should be disposed of accordingly . . . DON'T SELL THEM OR GIVE THEM AWAY. Destroy them by smashing the various items into bits, so that all traces of identification, serial numbers, etc., are oblitereated. Finally, it is recommended that you remove them some distance away and burn them with a good flammable liquid, which further destroys them and further obliterates all forms of identification. If you are in a position to avail yourself to the use of the local garbage dump or know of a large garbage dumster (such as is used by local businesses, restaurants, etc.), spread the burned and disfigured pieces through several of these. Ensure, again, that you leave no fingerprints on the bags or individual parts, or put in no other papers, etc., that might be traced back to you or an associate.

If you expect to be doing numerous future Black Bag jobs, and you have secure facilities for storage, such as a good summer hideway cabin and land for privacy, you should have one or several well-concealed hiding holes within a reasonable distance thereof — not immediately in or around the cabin itself.

When you cannot, or do not avial yourself of something like the above, you must be prepared to create a hideaway place in your own home. The following is one such type hiding place that has worked well in the past, but like any hiding place, may never be totally foolproof:

The accomplished Black Bag operator is usually a person of diverse skills and knowledge, and many times, has a hobby that can be of great value in Black Bag work. Thus, consider a well constructed workbench in a corner of your basement. But, before putting up this workbench, you knock a hole in the wall (wood, cement, etc.), remove the dirt behind the hole, and put up a concealed entranceway. Even behind the workbench area, the door should be concealed. This can be accomplished by repainting, putting in false bricks and/or stonework, covering the entire area with a wood paneling or, like I did in my previous house, used the wall adjacent as a paint storage area and the wall, itself, as a testing place for paint. Thus, the wall [and entrance door] was effectively concealed from cursory viewers by several layers of 'test' paint splotches covering the entire wall area — running, dripping, and general paint splotches tend to cover up seams and cracks; and those that are visible are put down to numerous layers of cracked paint. After the door has been built and installed, then construct the workbench, and give it a semi-solid look by use of back-framing the bench with 1/8 " pegboard (for tools, naturally) or 1/4" plywood.

I seriously consider the view of one individual who paneled the entire basement, but used rough paneling (3/4" plywood in the workbench area, and hung all his tools that were possible on this plywood area. Even when the bench was removed, all that was seen was the rough paneling. By removing one of the entire 3/4" panels (it was set in on concealed hinges), the concealed door was then available for opening. When he moved from the area, he filled up the wall again with good cement, and put in good paneling along the whole area so it looked like the rest of the basement. The particular corner was reconverted into a small bedroom, complete with built-in closets (in front of the area where the secret door compartment formerly stood). He also used a very expensive wood, being secure in the knowledge that it would not be removed, due to its cost. How right he was when, several years later he had the opportunity to check out the old homestead (for other than reasons of pleasure).

A final area to consider after the operation is to ensure that your associates have been paid off. If the payoff is cash, make sure the sum is such, and the bills are old and worn — not new. Hopefully, you will be able to further trust these people — remember they now know something about you that they can use to bargain with if they get into a 'jam.'

Selection of Partners

In a way, this can seem like a 'closed subject area since some individuals who perform Black Bag jobs are totally free-lance and do not take on partners. Nevertheless, the time may come — or it will be necessary — to have one or several partners for their specialized abilities, be they safe experts, or a master of disguise and a 'silver tongue' wherein they may be used as an accomplice to gain entrance into the offices and learn specifics necessary in order to complete your operational planning, or be an expert in other specialized areas.

Having a staff or even a single associate is both a help and a hinderance.

As a help, they can do much of the legwork (running) for you, picking up certain items of equipment, performing time checks and general ground surveys of the immediate and surrounding area, take preliminary looks into the subject area proper (inside the building/office/home) under a variety of pretexts, contact other individuals who are specialists in certain fields, and also obtain certain pieces of information that you, yourself, are unable to obtain. They also provide a buffer between you and others whom you shouldn't come into contact with, thus assisting and assuring that you can later claim that you have never been with, or even seen X person or offices.

As a hinderance, the more people you have working for you, or know about what you do (in general or the upcoming operation) poses a threat to your own security. The more people that know about it, the greater the chances are that someone will let something slip, or brag just a little, about what is about to happen, [either before, during, or even after the job has been completed]. Next, they have to be paid, either on a flat rate scale, plus a retainer to hold them over until a job is completed, or a percentage cut of the entire proceeds that you will recieve for the information you obtain. Money talks, and a person with extra cash spends it; on the other hand, a person will want his payoff immediately after an operation, and you might not be able to supply it immediately, since you have not yet been paid for the information acquired.

Since they have known you for a period of time, and know what you do, and one or more Black Bag operations you have been associated with, your security falls down even further. At a later date, they may be in a 'bind' with certain authorities and make a deal — YOU — in return for a reduced charge, or even swap the information about you for other information that they can use to their advantage. In the former case, you are going to be receiving some free room and board for a while, but in the latter case, the individual still knows something about you which he can still use to your detriment later on. Thus, unless you are fairly good at disguises and keep your true identity a secret even from your associates, you have a security problem (and a survival one, as well).

The selection of one or more individuals to assist you should be considered with the greatest of care.

Know or learn about their background, 'qualifications' and training, muchless experience in this and related fields that may be of value to you. [Also, this information can be used against them at a later date, if need be — RULE 1 — PROTECT THYSELF].

When considering a person, look at him for either a one-time or long-range professional association. If the operation is such that it will definitley have repercussions in one form or another, consider using a disguise constantly with your associate and having the person for only a one-time deal; to keep such an individual on, unless totally trustworthy and one you have had professional dealings with before and over a long period of time, can prove fatal. The less known about you, where you live, etc., the better. RULE 2 — LET

NOTHING ABOUT YOUSELF BE LEARNED BY OTHERS. This includes letting him know about your car, city you live in, professional interests, specialized hobbies, etc.

Definitely consider talking/seeing the individual when in a disguise (heavy moustache and full face beard, different name, limp, glasses, etc.) so that they won't recognize you again in your 'normal' appearance.

Does the individual have hobbies or certain traits that can work against them; what traits/hobbies do they have that can work to your advantage. Are they skilled in photography, safe work, an excellent driver or lookout man, etc.

Can he be relied upon — totally — in a tense or critical situation.

Is he on time? Black Bag operations have been known to fail for want of a man taking a break, coming in ten minutes late, or not being at the right place at the right time for the exiting from the operational area with the vehicle.

Observe him walking, running, standing, talking. What expressions or movements does he have that are noticeable. Can he blend in with a crowd. What about physical or possibly mental abnormalities. Any specific habits that draw attention to himself. Does he wear flashy or 'seedy' clothes no matter where he is, and no matter what he is doing. Can he fit into different stratas of society. What about his speech patterns. Is there a peculiar accent, drawl, stuttering, etc.

All the above and more you should know before taking on an individual as an associate.

* * * * * * * * *

The determined Black Bag operator will see that in order to insure proper security and a timely entrance and exit on a Black Bag operation, research and planning in a must. Planning that has gone on previously now moves into extremely high gear as you make your final selection of 'associates', if any, who will be accompanying you on the Black Bag job.

The relative success of the Black Bag operation is based on the thorough coordination and flawless operating practices of all personnel involved.

The ability to select the individuals is, though, a subjective one. But in these type cases wherein professional expertise is so necessary and valuable, objectivity and what an individual can contribute towards the successful completion of the operation is the "MUST" that is the final consideration when taking them into your confidence as an associate.

The personnel that could be involved in various types of Black Bag operations include:

— An organization/coordination expert (not necessarily you) who supervises the various phases of planning, preparation, determining and obtaining supplies and equipment or arranges ot obtain them, hires and pays the individuals (one time) who are involved, and who may or may not know what specific information you are looking for on the Black Bag job. He usually accompanies other individuals on the operation, but stays in the background, possibly doubling as a communications man or lookout. He has overall knowledge of all phases of the entire Black Bag operation.

— A lock and safe expert/locksmith. He is knowledgeable in the operation and techniques for the rapid opening or bypassing of many types of door, window, and padlocks. He is also accomplished in the art of opening safes by various manipulation techniques. Either by hand or by machine he can quickly duplicate, accurately, keys for various types of locks.

— A camera expert who can produce excellent quality products with a minimal amount of equipment, and who can work under varying conditions and restrictions which include setting up and dismantling equipment in less than normal time; requires a very small work area for photographing documents, equipment, etc; can make excellent photographs and enlarge them. The individual selected should have several miniature and standard camera and the ability to use light-sensitive film, muchless infrared, and other types.

— A 'flaps and seals' expert who is technically qualified in the opening and resealing of various types of envelopes and wax seals, and packages. He must be extremely skilled and knowledgeable, and able to work under extreme pressure without losing his temper or making any mistakes, and under extremely short time constraints. Also when his work is completed, examination of the letter, package or wax seal should not indicate that even an attempt was made to tamper with it.

Lookout(s). One or several may be employed. They also may have other collateral duties involved while on lookout duty. These individuals take pre-positioned areas for short- or long-term periods of time to ensure the security of the person(s) actually performing the Black Bag operation. They must be 'plain-looking', not noticeable, and able to go for periods of time doing almost absolutely nothing, but still be able to put up a front against inquisitive strangers or several general questions.

— Communications Expert. At least one person, who is extremely knowledgeable about radio communications, and who can monitor local police frequencies, and other possible frequencies of private investigative/police forces, and also the CB radio channels, and who is in open contact at all times with all other members of the operation. He should know where to obtain or be able to build unique type pieces of equipment that may be necessary to the operation, such as short-term miniature transmitters for emplacement in the area of operations for future use, and have available certain jamming equipment modules, if necessary, to jam other sensitive communications (such as police/security force). He should also be aware of, and able to perform, telephone monitoring services to the lines used by local security guards, if necessary, and able to do this in such a way that defies normal detection.

When possible, his equipment should be in modular equipment cases, such as suitcases, with each one self-contained and battery-operated. Equipment he purchases or builds should not be traceable.

This Page Is
Intentionally Blank

Operational Clothing Requirements

The selection of type, style, color and uses of clothing worn on operations is dependent, to some extent, upon the area and locale, muchless the time of day or night and the type of area you are working in (working area of lower middle class people, exclusive 'wall street' business area, run-of-the-mill department store, etc). The initial criteria is determined by the type of clothing to be worn into and away from the site of the operation and any specifics that may be required, such as unique clothing requirements [disguises] such as delivery man uniform, rough working man's clothes [to include dirt and other stains], professional business suit, etc.

When working at night you should have a double set of clothing available, not necessarily one worn over the other. At times you may want the first set to conceal the second set, this being for the purpose of either entering or exiting the general area without seeming out of place with the rest of the general populace. In any case, the outer clothing should be subdued, conserative, in the style that is current, and within the approximate price range of others appropriatley dressed so that you 'blend' in with other individuals within the general operational area.

Operational clothing, during daylight hours, will be developed normally and will — usually — be the outer set of clothing, unless entering a sensitive or exclusively denied area in which general clothing cannot be worn. In such cases, the specialized clothing should be obtained and either secreted within the operating area several hours to a day or more in advance, or else worn under your outer garments (in some cases, such items as a 'company coat' or 'white lab jacket' can be conveniently carried under your arm — rolled up, in a paper sack, or in the briefcase).

During night operations, when concealment is a must, the clothing will be a dull black, with no metallic fastenings (buttons or velcro fastening to replace zippers, etc). Clothing should be well fitting, but not overly tight, comfortable, and have a number of loose pockets on the exterior of the clothes with closing covers with small strips of velcro attached so that items within will not fall out of pockets.

A black sweater is preferable as a top, but it can have drawbacks. You don't normally wear such during a hot summer and an overprofusion of perspiration and general itchiness can result. A sweater stretches when necessary, and its elasticity can help to hold a small wrap-aroung carry-all pouch belt that can be tucked underneath.

Small and meduim-size pockets can be made from black material and securely affixed to the upper arms, the upper and middle front portions for carrying certain items that you may need immediately. Such items will always be lightweight, and when possible, buy only items that the specific individual will use; no one else.

A Navy-type watch cap or similar type of design should be used for the head, It should be a quick-pullover type and be close-fitting. It will cover the hair, back of head, and possibly the ears. It should not be so tight as to make wearing a radio earpiece uncomfortable.

Ultimatley, all clothing should be chosen for their ability to absorb rather than reflect light, either direct or indirect.

When considering shoes for operational use, two types are considered. In general, light athletic shoes, colored black, are quite acceptable. Metal eyelets should be removed and special sewing accomplished around the string holes. This can also be done by using a heavy webbing material over the top (one piece to each side) and folding it down and under to cover the inside areas. After complete sewing is done, a light covering of an absorbant glue can be used (Elmer's glue and water: 1-1 proportions). When this is dry, cut a very small hole where the eyelet hole is. The lacing for the shoes will contain no metal tips.

Soles for these shoes or other types should be of the 'crepe' type. This particular type give the best grip for short climbs, and lesser amount of 'squeek' noise when you are walking. When used, they should not be wet or damp; this will account for much of the 'squeek' heard. Make sure, also, they have been broken in properly and are extremely comfortable .

The second type of shoes are 'friction boots.' They are probably the best you may find for use in certain situations wherein climbing may be a 'must,' such as entry or exiting from a second-story, or higher, or a building. They are extremely thin and lightweight and have been developed to provide for the maximum amount of grip and temporary adherence to an object, such as walls or rocks, etc. While initially developed for mountain-climbing where going up sheer or near-sheer surface rockfaces where there is little or no general areas for a regular shoe or boot to hold in place, the 'friction' portion of the soles and tips are absolutely fantastic in their ability to cling where you would not think it possible. They can be obtained from most stores that sell mountain-climbing equipment and supplies. Prices are reasonable for these items.

General Notes

If a sweater is not to be used, then a loose-fitting shirt with long sleeves should be considered. If so, try and make it a reversable one, with the reverse side a different color/pattern than the operational side. In cases such as this, checkered, striped, etc., have been used successfully. Immediately after exiting the site, the reverse side is worn. Later it can be changed back in a different sector of town, thus totally alternating your outward appearance [also, they are — or might be — looking for someone with a different colored shirt] .

Sox should be slightly long so they can tuck trouser legs into them. This avoids the possibility of the leg material getting caught on something.

Sleeves should be somewhat close-fitting on shirts.

Use of velcro strips for wrap-around fastenings is excellent for trouser legs, sleeve openings, etc., when material may be overly loose or flowing. Always use velcro for pocket flap closures, vice buttons or snaps.

I have found that standard racing car driver type gloves are excellent; while somewhat tight-fitting, they have the air ventilation holes, and are easy to put on and remove. They also take up less space than other gloves. They are well-made, thin but sturdy, and give an excellent grip without loss of muscle tension. If they are not black, dye them in a heavy strength dye.

Equipment and Assorted Paraphernalia

Two types of equipment are considered 'necessary to the Black Bag operation: that used in the planning phase, and that which will be used during the execution of the Black Bag operation. Some of the planning phase equipment may also be considered backup or emergency type to be considered for the execution phase, should the need for it arise.

Basically, equipment is kept to a minimun, is small and as light-weight as possible. Whenever possible, also, it should all be 'legal' and not of the type that might arouse suspicion by the ordinary person when seen out of context, i.e., late at night, during peroids of darkness or on weekends, in areas where it would not normally be. Also, it should be considered that if a person carrying such equipment is stopped by the police, the equipment is innoculous enough and a proper statement concerning it is presented, the police would not further question you about it, or take on other suspicions about you or the equipment.

Considerations that must be seriously studied when selecting and/or determining the type, nature, and amount of equipment to be used in a Black Bag operation include, but are not limited to:

— Cost

— Is it a peripheral piece of equipment or actually necessary to the operation; is it "nice to have" — items that are not really essential for the operation or in maintaining appropriate cover should be avoided.

— What is its utilitarian use.

— What is the weight of the object.

— What is the size (overall dimensions); will the item fit comfortable into a briefcase or other innoculous carrying device without seeming too conspicuous or draw attention to itself or its carrier. If it is overly large or unique (such as a pole ladder), can it be further disguised as another object, or else redesigned and configured to be broken down into several pieces and quickly assembled at the scene of the operation.

— Can it be discarded easily afterwards (consider cost and relative positive merits when considering item for purchase or construction).

— Is the item of such a type that it can be traced back to you as the purchaser or current owner. Grappling hooks, for instances, are not sold in such an amount to a wide and varying segment of the population. Also, they are not found in many sectors of the country, sitting out on a hardware store's shelf; consider the source and where you are located when considering the product.

— Is the item unique/unusual in any way (example: some types of radios purchased by the 'hobbyists' are okay; but when one person comes into a single store and purchases an amount that would not normally either be purchased by one person or else sold to several over a period of several months, the owner tends to remember — and recognize — you. When located later as a 'junked' item, the owner might well come forward and say . . . . [REMEMBER SECURITY — THINK OF NUMBER ONE!]

— Have a plausible excuse for buying any item that you need. Also, never overbuy or underbuy.

— In the instances of certain types of small radio transceivers or one-way units, police scanners, etc., when time permits, consider having the electronics expert purchase the parts from a radio supply store (Radio Shack, etc) and then build the equipment. While it is more time-consuming, it leaves no trace of the individual purchaser, and the equipment can be configured to a specific design or module.

— Don't make all purchases of many unrelated items in a single store. Spread them out. Try to relate several items and purchase all of one and part of another at one store, and then move on to another store to pick up the rest of the parts for the second item, while purchasing a portion or all of the third item needed. Try to relate one or several items in a single package. Also, you can mislead a salesperson as to the intended use by buying immaterial and totally unrelated items that you will shortly discard at a distant location.

— PAY CASH — DON'T EVER CHARGE FOR BLACK BAG EQUIPMENT!!!

— Men: when purchasing cloth for blackout curtains, be careful; you probably aren't really as knowledgeable about cloth as a woman would be. Use a plausible excuse for this and other like items that you would normally be unfamiliar with. I have found that by representing myself as a worker from a local or nearby church or civic group that is doing community theater, they are willing to be of more assistance, while not being overly inquisitive about its actual intended use. You should have a plausible cover story, i.e., the play and something about it, in mind, to carry on a line of conversation. Also, the type and quantity of material is never too excessive, since they think you may well be coming back for more and need only XXX amount for now.

The above, and other types of cover stories and ruses are excellent groundwork and training for building up more potential cover stories to fit in with the pre-operational details that may require an accomplice or yourself getting onto the premises to 'scout' out the situation.

* * * * * * * * *

No list of supplies and equipment for use in Black Bag operations can ever be considered all-inclusive without seeming overweighted with unnecessary items, or underdeveloped with many items left out. This is due to the fact that what one Black Bag individual sees as essential, another sees as either non-essential or peripheral to an operation. Therefore, included here are basic and essential items of equipment necessary for use on, or in developing specialized pieces of equipment that would be used during Black Bag operations. Depending upon exactly what is required in any given operation,the amount, number, type(s), and quality of items of equipment will vary. Also, certain items that might be used on the first Black Bag operation might never be used for ten years or more, but it could always be considered for futrue use. By the same token, what is set aside as possibly a supportive piece of equipment for a number of operations that are performed over a period of weeks or months.

Tools. Basic handtools are not always essential — directly — to an operation, but are always considered peripheral. They can be used, and often are, in creating specialized equipment. Thus, you should have a good workbench area and a wide variety of handtools and metalworking equipment for current and future work needs.

Certain tools that might be taken on a Black Bag job can include: pliers (various types), hand or power drill and bits, rubber mallot, claw hammer, regular and miniature screwdriver sets, etc.

Cable/Rope. Nylon cable, 1/8" to 3/16" diameter can be used for pull-over and carrying objects between buildings and up/down sides of buildings. Also, good for supporting personnel between these points. Airplane cable 1/8" diameter is more than adequate for supporting an individual and one or two carrying cases at one time.

When using these, especially the airplane cable, you will want small "U" clamps and a miniature wrench to tighten the clamp nuts down to secure the cable. All these items are small and very compact and can easily be carried in a coat or jacket pocket.

— Grappling Hook; can either be purchased or of own design. Consider weight and concealment factors. Used for end weight on a throwing line (cable or rope) when either scaling side of building or crossing over between two points (over alleyway between two buildings).

— Cable wheel set (2 preferred; one for backup); should be constructed to own specifications for various operations. Used to move across cable — either individual and/or equipment cases.

— Pole climbing straps; Used by individuals for climbing up/down poles, drain spouts, etc., for entrance/exit from second-stories or higher.

— Equipment cases; consider using standard briefcases or variations thereof to fit particular job.

— Carry-all pouches; worn over/under shirt/sweater, depending upon bulkiness of items carried. Attached with velcro and/or hook snaps. Design is up to individual, but dependent upon type and amount of small equipment to be concealed, but carried on person.

— Lockpick set; consists of standard picks for warded and cylinder locks and tension wrenches and several small files and key blanks.

— Electronic stethoscope. Comes with earplug and can be used for: checking out adjacent room prior to entering (ensures room is empty); extremely useful for working on small safes in opening them by manipulation).

— One-way or two-way transmitter/receiver units. One-way units usually for alerting members of team on inside by individual on outside of impending danger. Two-way unit usually between supporting members and one member of inside team.

* * * * * * * * *

Equipment carried by individuals on penetration team should be in standard type containers, usually briefcases or small bags, depending upon amount of equipment necessary. In office buildings, standard briefcases should be used — and are preferable — since they assist in creating an aura of authority and professionalism associated with the white-collar worker. Also, they lend a certain amount of respectability to the carrier's presence while approaching or leaving the operational site. If other types of cases are carried, they would tend to draw more attention than a standard briefcase.

During later hours of darkness, basic materials can be carried in haversacks or similar cases.

Equipment that is carried in briefcases might have a company name put on since these provide addtitonal 'cover' support. Each individual should carry their own equipment plus, possibly, certain pieces of initial equipment necessary in order to obtain the proper entry/exiting from the area.

When putting items into briefcases, they should be placed in order of: first item in = last item used; last item in = first item used.

In many instances, the Black Bag operaton will be performed by only one person, which necessitates him carrying all equipment and associated materials. When a lookout is not even considered, the indivudual will probably pre-position certain items, concealing them on rooftops, etc., ready for use, but then removing them when he departs the operational site area.

The carry-all pouch is a close fitting vest type apparatus or can be a variation thereof. The basic criteria for this is that it not contain too many items required, and that the items be small, compact, and light-weight. No item should protrude from the small pockets. For best results, pockets should have a flap covering and a small tab of velcro to hold them closed at all times. No pocket should have more than one type item in it.

For the best construction use a vest or wide cloth like that of a cumberbun (from a dress tux), affix the various items and mark the outside dimensions. leaving an extra 3/8" to 1/2" for ease in inserting/removing items. Cut appropriate cloth and attach. Make a pocket flap and sew on, attaching a small tab of velcro to the flap for securing the pocket in a closed position.

Don't put too many items on the vest/cloth strip and overload it, and keep items spaced apart.

Consider the use of a telephone lineman's belt, surplus Army web belt or double thickness long leather belt. Whatever you decide to use, you should have clip hook fasteners or large "D" rings that can either be quickly attached or are a part of the belt unit. The belt serves as a holder for pieces of equipment that you may want to carry with you, but are too large for a carry-all vest piece, or equipment that you may need for exiting or entering, such as over roof tops (rope, grapple hook, etc). Belt would normally be discarded once entry is affected and not picked up until the exiting process begins.

Cover props may be necessary for a proper Black Bag job, in order to give adequate support to a 'cover' story. Building sway test equipment, telephone company equipment, building engineers, electricians, etc., all have specialized pieces of general equipment they carry and use in performing their various job functions. If it is necessary to disguise yourself such a fashion, remember that the clipboards, notes, dress, tools and other equipment should look used, and you should be able to act the part, and also discuss various technical phases of a job with an individual, if need be.

In considering cover props, remember that a device made as a cover prop may seem to perform one function when, in reality, it actually performs another. Electronic equipment cases and devices can have the inside parts removed and other parts included, such as one-way or two-way transmitter unit inside a voltmeter case, or a police scanner unit inside a VOM or line-testing voltage meter box.

Along with the general pieces of equipment needed for a Black Bag operation, you should have the following items:

— Flashlights: small penlight flash or else the miniature flashlights that are avialable for under $1.50 in local drugstores; these are a throw-away variety so you can dispose of them quickly. Unless the flashlight is new, have at least one extra available for use.

— Sap (blackjack): for use in subduing individuals who may cause trouble or give alarm; this should not be confused with methods for rendering persons immobile when they are found in or near your operational area. For these, you should use the following item.

— Emergency Interruption Kit: consists of several short rope lengths, adhesive tape (2" wide — athletic tape), gags and blindfolds. Cotton stuffing may be used for ears of individuals.

— Blackout curtains: cut to rough size of windows but slightly larger for overlap with window frame and wall.

— Roll of tape, 1" width: can use adhesive from interruption kit for this; purpose is to hold up blackout curtains, tape back (temporarily, door locks, etc).

— Flaps and Seals Kit: consists of several wooden knives, 'magic' tape, tweezers, etc.

— Key cutting files, key blanks, etc.

While weaponry is not, and should not be condoned during Black Bag operations, gas guns, stun guns etc. are acceptable. The Taser gun is an an excellent method for quick immobilization of possible opponents. A taser gun can be slightly modified to have extra (duplicate) wires, and you should consider the possibility of modifying further the Taser so that two 'shots' can be fired at two different opponents in quick succession.

This Page Is
Intentionally Blank

Appendix A

Information Obtainable From Black Bag Operation

In general, a wide variety of information type items can be obtained from Black Bag operations. They can include, dependent upon the specific nature of the operation and where you look for the information:

- Employee and membership listings
- Photographs of individuals (including security and personnel type pictures; usually in files)
- Business relations with other businesses
- Current and future business actions and activities
- General and specific items of correspondence
- "Academic" debates/reviews/papers relating to certain subjects and issues of general/specific interest
- Civil rights
- Anti-administration, advertising, research and development paperwork
- Inter- and intra-office memorandae that is normally closely controlled
- Specific files (personal) held by senior and middle management officials
- Specific file listings on various subjects [possible future Black Bag operational targets]
- Travel planning for conferences, etc. [possible future Black Bag operational targets]
- Short- and long-range future planning
- Current status of various projects

— Proposed projects and various stages of planning (different from general short/long-range planning)

— Publishing plans

— Financial Status

— International travels

— Legal defense strategy

— Contract negotiations and secret aggreements thereto

— Political contacts and 'in' persons, to include possible political conflicts

— Procedural planning books and manuals

— Company proprietary 'secrets' (formulas, processes, etc)

— Letterhead stationary (blank), reproduced stationary, and secondary sheets and envelopes. Reproduced stationary would/may have appropriate signatures attached

— Signature blocks and organizational charts

— Company directories

— Specialized equipment listings — on-hand or on order

Depending upon the type of business the organization is engaged in, the above list may be added to. Some initial Black Bag operations are, initially, fishing expeditions, just to ascertain if there is possible information that can be obtained at a future time(s). Any and all information is noted during these operations, and then it is filed away for future use.

Only time constraints and operational security, muchless the ability of the personnel involved with the operation, can tell how much useful and peripheral information can be obtained and used immediately, or set aside for future operations or use.

Appendix B

Sample Plan of Action F. B. I.

To give an idication how planning can be detailed in a report in preparation for a Black Bag operation (surreptitious entry in FBI jargon), the following is extracted from an FBI memorandum which contained detailed plans for a post-midnight burlarizing of an office building headquarters. The entry plan included descriptions of 'security aspects' such as building floor plans, locks, lighting, surrounding streets, entrances, and the occupants' living habits.

> "Several memoranda describe the 'security aspects' on the FBI agents' plans for security entry into the headquarters . . . . One reads as follows:
>
> "The headquarters entrance is a store front on the street level. There is only one entrance to the headquarters. The door is locked with a Master padlock only . . . .
>
> "The entrance to the building is located approximately 75 feet on the north side of [the] street from Second Avenue. The headquarters is a street front located adjacent to the entrance to the apartment building . . . . East of the headquarters store front are located four similar store fronts within the same building. These are described as follows from the headquarters going East: New York Telephone Company; empty store front; law office; empty store front.
>
> "There are 4 floors of apartment dwellings above these store fronts in the building.
>
> "There is a street light located on the north side of [the] street, approximately five store fronts east of the headquarters. Inasmuch as the nearest other street light is located on the southeast corner

of [the street] and Second [street], the immediate area of the headquarters is reasonably dark in evening hours.

"Previous spot checks on numerous occassions have shown that there is a very limited amount of pedestrian and automobile traffic after 12 midnight. These spot checks have also shown that the light of the apartments in the building are darkened.

"Entrance will be made between the hours of 12 midnight and 4 AM, [date]."

When the headquarters moved, the 'security aspects' of the FBI's entry plans were reevaluated, as follows:

"This building is a three-story edifice approximately 25 feet wide by 75 in depth. The second and third floors are loft premises. The first floor is occupied by [a paint company]. The entrance to the second and third floors of the building is a door located beside the paint store. This door leads directly to stair flights to the second and third floors and is secured by a cylinder lock. This entrance does not connect with the paint store on the street level . . . .

"The third floor loft of this building is occupied by an artist . . . who maintains a studio. This individual pursues his profession, together with holding occassional art classes, in this loft. This activity transpires during the daytime. [The artist] does not reside on these premises and is not known to frequent the premises in the evening hours.

"The [deleted] headquarters are located on the second floor loft space. The [deleted] moved into these headquarters on [date]. Numerous spot checks of the area have shown very limited pedestrian and automobile traffic after midnight. The building adjacent to this location . . . on both sides of the street, are commercial establishments and lofts, and contain no residences.

"It has been ascertained that the paint store at this building close business between 5:00 and 6:00 p.m . . . .

"Entrance will be made between the hours

of twelve midnight and 4:00 a.m., on [date]."

From the above, you can observe that the planning stage involves a number of details. Essentially, the legwork involved took numerous hours to check, and the final report was short, concise, and covered all the basic details that were necessary in order to ascertain the general layout of the target building and surrounding area. Naturally, other more detailed plans would have been drawn up if it were necessary to have more information about the subject building and its occupants. In this particular case, the background and motives of the occupants were already known to the FBI, thus it was not necessary to include them in the basic plan, since they did not keep irregular hours, have alarm systems present, etc., in the subject building.

This Page Is
Intentionally Blank

Appendix C

References

A number of general and specialized books are available to the individual who needs assistance in developing his Black Bag expertise in various areas. The following books and manuals are available from the publisher of this book:

Bugs & Electronic Surveillance

CIA Flaps and Seals Manual

Combination Lock Principles

Electronic Stethoscope Plans

Handcuffs I

How to Fit Keys by Impressioning

Keys to Understanding Tubular Locks

Lockout — Techniques of Forced Entry

Lock Pick Design Manual

Lock Picking Simplified

Locks, Picks, & Clicks

Master Key Locksmithing (Vols I & II)

Physical Security

Police Guide to Lock Picking & Improvised Lock Picks

Safe and Vault Manual

Techniques of Safe & Vault Manipulation

The Plumbers Kitchen

Wiretapping Handbook

Aside from these, other subject areas you should consider include books on espionage, police, crime, etc. Many technical books, under other innoculous titles exist, but if you don't know what you are looking for, you'll never find it.

Along the way to building an excellent library, don't forget that general notes and ideas that you generate or obtain require you to delve further into related subject areas in order to 'create' a specialized piece of general or special electronic or other type equipment. So, after a while, you will have nice couple of shelves of books that will not draw any attention to you, since they will cover an extremely wide variety of subject areas.

When considering ideas and techniques, watch the police, detective, and related type crime shows on television. The techniques and equipment they use can, at times, be very excellent ideas for a future Black Bag job. When you see an item, jot down its name or a physical description and a quick sketch. You may see the item while shopping some day and suddenly realize its potential.

Hardware stores, in order to order the various pieces of hardware they sell, have very interesting manuals that you do not normally see, and are not normally available to the average person outside the hardware business. Drop by and see if you can get one or several old hardware catalogs; they cover a myriad of tools, equipment, parts, and lots of information about the uses for the products. The ideas the book can give you are tremendous.

A general magazine, while not specific in details, has quite a number of advertisements that can be of interest to the average "Black Bagger." Popular Science, Popular Mechanics Illustrated, True, etc. All these, even the vintage issues of twenty years ago, are excellent references, when you consider the advertisements. Great ideas come from a simple advertisement. Also, consider Soldier of Fortune magazine. A professional magazine for mercenaries, it offers products, listing, ideas, etc., that are absolutely fantastic and that can be used by the Black Bag individual who needs a little piece of self-protection or specialized knowledge in order to carry out an operation.

Notes